Name_____

Numerators and Denominators

Faye <u>divides</u> things into equal parts.
They are the <u>denominators</u>.
Write the denominator under each picture.

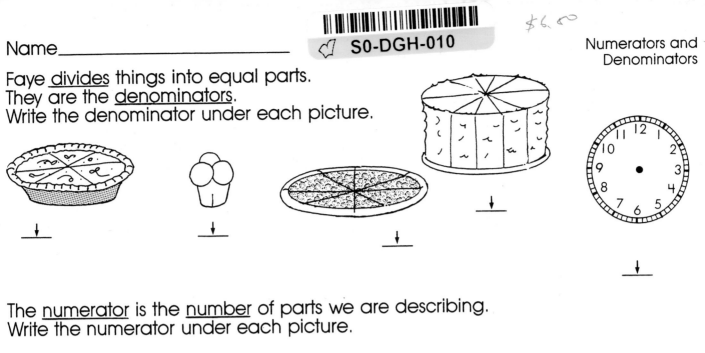

The <u>numerator</u> is the <u>number</u> of parts we are describing.
Write the numerator under each picture.

Fernando ate _____

Faye ate _____

Fred ate _____

How much time? _____

Frieda ate _____

Write the numerator and denominator for each drawing below.

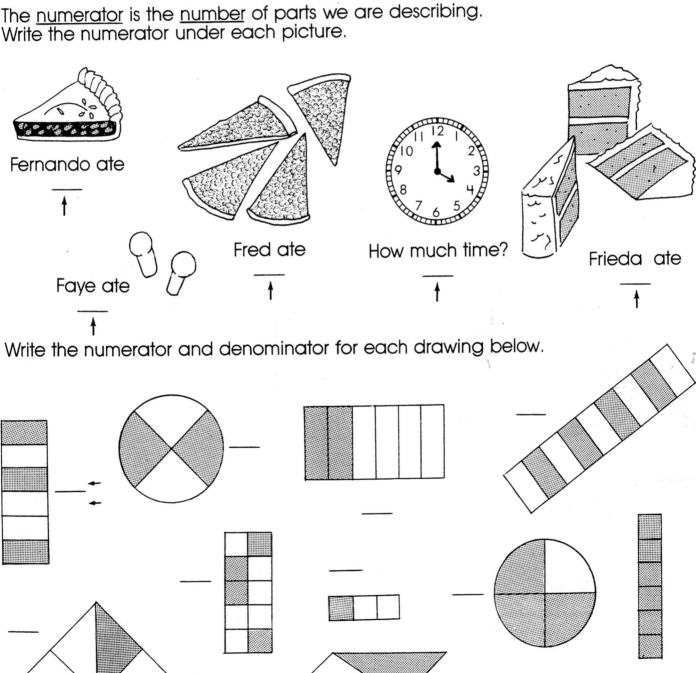

_____ or _____

© Milliken Publishing Company

MP3378

Name: Jonathan

Seeing Fractions

Fred was the marble champion!

When he left for school, he had 12 marbles in his bag. Fred laid them out in 3 rows with 4 marbles in each row.

Frieda wanted to borrow $\frac{1}{3}$ of the marbles. Draw a line through $\frac{1}{3}$ of Fred's marbles. You may write this as a division sentence:

$$12 \div 3 = 4 \quad \text{or} \quad 3\overline{)12}^{\,4}$$

Write each of these as a division sentence.

$\frac{1}{3}$ of 9   3       $\frac{1}{4}$ of 12   3       $\frac{1}{3}$ of 6   2       $\frac{1}{2}$ of 10   5

Now, write a division sentence and the answer for each problem below.

1. $\frac{1}{2}$ of 10    5    2. $\frac{1}{3}$ of 9    3. $\frac{1}{6}$ of 12    4. $\frac{1}{5}$ of 20

5. $\frac{1}{6}$ of 30    6. $\frac{1}{8}$ of 64    7. $\frac{1}{7}$ of 21    8. $\frac{1}{12}$ of 24

9. $\frac{1}{2}$ of 80    10. $\frac{1}{10}$ of 50    11. $\frac{1}{3}$ of 66    12. $\frac{1}{10}$ of 100

13. $\frac{1}{5}$ of 25    14. $\frac{1}{8}$ of 32    15. $\frac{1}{7}$ of 49

Total Correct
What fraction is that? ⎯/15

© Milliken Publishing Company

Name_____     Equivalent Fractions

Frieda was hungry. Faye left a cake cut into 2 pieces. Frieda ate $\frac{1}{2}$ of the cake.

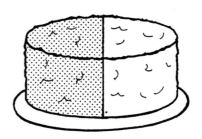

If Faye had cut the same cake into 4 pieces, she would have eaten $\frac{2}{4}$ of the cake. $\frac{1}{2}$ and $\frac{2}{4}$ are the same or equal fractions.

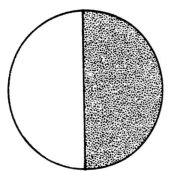 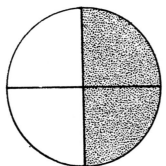 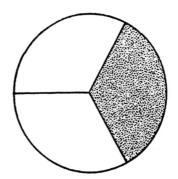 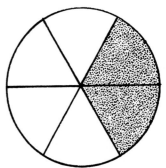

$\frac{1}{3}$ and $\frac{2}{6}$ are also equal fractions.

Match the equal fractions below.

1. $\frac{6}{8}$  $\frac{1}{2}$      2. $\frac{4}{10}$  $\frac{1}{4}$      3. $\frac{2}{3}$  $\frac{6}{9}$      4. $\frac{2}{5}$  $\frac{1}{2}$

   $\frac{5}{10}$  $\frac{2}{3}$        $\frac{3}{12}$  $\frac{2}{5}$        $\frac{5}{15}$  $\frac{3}{15}$        $\frac{2}{3}$  $\frac{4}{5}$

   $\frac{6}{9}$  $\frac{3}{4}$         $\frac{8}{16}$  $\frac{1}{2}$         $\frac{4}{8}$  $\frac{1}{2}$         $\frac{18}{36}$  $\frac{4}{10}$

   $\frac{4}{12}$  $\frac{1}{3}$        $\frac{1}{2}$  $\frac{1}{7}$          $\frac{1}{5}$  $\frac{1}{3}$          $\frac{8}{10}$  $\frac{4}{6}$

   $\frac{1}{4}$  $\frac{2}{8}$         $\frac{2}{14}$  $\frac{3}{6}$         $\frac{14}{16}$  $\frac{7}{8}$         $\frac{3}{4}$  $\frac{12}{16}$

Write your number correct as a fraction.

© Milliken Publishing Company                3                              MP3378

Name_____

>, =, <

1.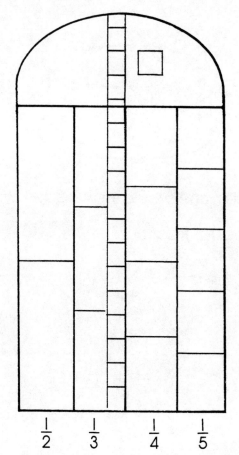

$\frac{1}{2}$  $\frac{1}{3}$  $\frac{1}{4}$  $\frac{1}{5}$

2.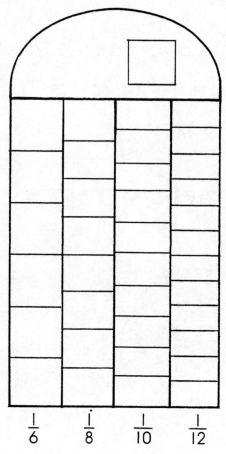

$\frac{1}{6}$  $\frac{1}{8}$  $\frac{1}{10}$  $\frac{1}{12}$

If silo #1 is $\frac{2}{3}$ full of feed and silo #2 is $\frac{3}{8}$ full, which has the most feed, #1 or #2? _____

Which is greater >, equal =, or less than <? Put in the correct sign.

1. $\frac{1}{2}$   $\frac{1}{4}$
2. $\frac{1}{12}$   $\frac{1}{5}$
3. $\frac{1}{3}$   $\frac{1}{10}$
4. $\frac{1}{8}$   $\frac{1}{6}$
5. $\frac{2}{8}$   $\frac{5}{10}$

6. $\frac{6}{12}$   $\frac{3}{6}$
7. $\frac{10}{12}$   $\frac{3}{4}$
8. $\frac{1}{2}$   $\frac{4}{8}$
9. $\frac{1}{4}$   $\frac{6}{8}$
10. $\frac{1}{2}$   $\frac{4}{4}$

11. $\frac{2}{4}$   $\frac{5}{10}$
12. $\frac{5}{10}$   $\frac{6}{12}$
13. $\frac{3}{10}$   $\frac{5}{8}$
14. $\frac{3}{4}$   $\frac{5}{6}$
15. $\frac{1}{3}$   $\frac{1}{6}$

16. $\frac{4}{8}$   $\frac{5}{10}$
17. $\frac{2}{3}$   $\frac{2}{5}$
18. $\frac{6}{8}$   $\frac{7}{12}$
19. $\frac{9}{10}$   $\frac{11}{12}$
20. $\frac{7}{8}$   $\frac{6}{6}$

© Milliken Publishing Company

MP3378

Name_____     Adding Mixed Numbers

## ADD AND REDUCE

$\frac{1}{3} + \frac{1}{3} + \frac{1}{3} + \frac{1}{3} = \frac{4}{3}.$   $\frac{4}{3}$ is written as $1\frac{1}{3}$.   $1\frac{1}{3}$ is called a mixed number.

Add and reduce these fractions.

1. $\frac{4}{6} + \frac{6}{6} = \frac{10}{6}$ or —

2. $\frac{2}{8} + \frac{8}{8} = $ — or —

3. $\frac{3}{5} + \frac{3}{5} = $ — or —

4. $\frac{2}{3} + \frac{2}{3} = $ — or —

5. $\frac{5}{6} + \frac{3}{6} = $ — or —

6. $\frac{7}{5} + \frac{2}{5} = $ — or —

7. $\frac{2}{7} + \frac{9}{7} = $ — or —

8. $\frac{2}{3} + \frac{1}{3} = $ — or

9. $1\frac{1}{4} + 1\frac{1}{4} = $ — or —

10. $4\frac{2}{10} + 2\frac{1}{10} = $ — or —

Write four mixed number addition problems. Be certain the answers are greater than 1 for each problem.

11. $3\frac{1}{5} + 1\frac{3}{5} = $ — or —

12. $1\frac{5}{6} + \frac{2}{6} = $ — or —

17.

13. $1\frac{2}{3} + \frac{2}{3} = $ — or —

18.

14. $1\frac{5}{10} + 1\frac{1}{10} = $ — or —

19.

15. $4\frac{3}{8} + 1\frac{1}{8} = $ — or —

20.

16. $4\frac{2}{10} + 3\frac{8}{10} = $ — or

© Milliken Publishing Company

Name_____                                    Fractions as Decimals

## FROSTED FUDGE COOKIES

Everyone loves cookies. But, today, the cookies stand for <u>decimals</u>.
(Remember—that always means <u>10</u>.)

Write the decimal for each fraction shown below.

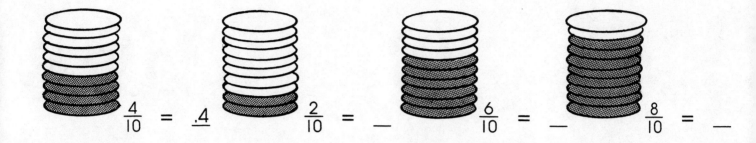

$\frac{4}{10}$ = .4      $\frac{2}{10}$ = ___      $\frac{6}{10}$ = ___      $\frac{8}{10}$ = ___

Write the missing numbers below.

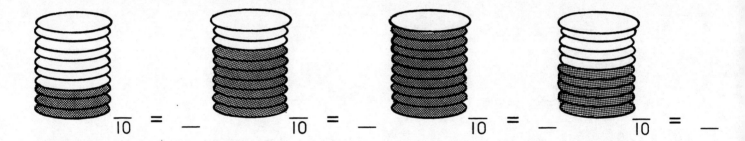

$\frac{\phantom{0}}{10}$ = ___      $\frac{\phantom{0}}{10}$ = ___      $\frac{\phantom{0}}{10}$ = ___      $\frac{\phantom{0}}{10}$ = ___

Write as decimals.

A.  $\frac{3}{10}$ = ___      $2\frac{1}{10}$ = ___      $4\frac{6}{10}$ = ___      $6\frac{4}{10}$ = ___      $\frac{5}{10}$ = ___

B.  $1\frac{3}{10}$ = ___      $7\frac{7}{10}$ = ___      $1\frac{2}{10}$ = ___      $4\frac{1}{10}$ = ___      $\frac{9}{10}$ = ___

Write these decimals as fractions. Reduce if possible.

C.  .7 = ___      .2 = ___      1.6 = ___      9.2 = ___      .4 = ___

Circle the larger number.

D.  .4 or $\frac{6}{10}$      $1\frac{5}{10}$ or 5.1      2.6 or $1\frac{9}{10}$      1.8 or $1\frac{7}{10}$

E.  .4 or $\frac{1}{2}$      1.3 or $\frac{11}{10}$      .3 or $3\frac{1}{10}$      .9 or $\frac{9}{100}$

Name_____  Subtract and Rename

## GET A LEG UP!

Subtract and rename.

$9\frac{3}{4} - 1\frac{1}{4}$

$2\frac{5}{6} - 2\frac{2}{6}$

$6\frac{2}{3} - 5\frac{1}{3}$

$7\frac{5}{6} - 6\frac{1}{6}$

$7\frac{6}{16} - 2\frac{2}{16}$

$7\frac{4}{5} - 4\frac{3}{5}$

$6\frac{4}{7} - 5\frac{2}{7}$

$9\frac{10}{12} - 7\frac{9}{12}$

$6\frac{3}{4} - 2\frac{2}{4}$

$8\frac{9}{10} - 7\frac{8}{10}$

$1\frac{3}{12} - \frac{9}{12}$

$8\frac{4}{8} - 2\frac{7}{8}$

$2\frac{1}{10} - \frac{6}{10}$

$10\frac{2}{10} - 9\frac{2}{10}$

$6\frac{5}{10} - 3\frac{7}{10}$

$4\frac{1}{7} - \frac{5}{7}$

$9\frac{3}{10} - 4\frac{6}{10}$

$1\frac{1}{8} - \frac{7}{8}$

$8\frac{4}{5} - 3\frac{1}{5}$

$1\frac{9}{12} - \frac{3}{12}$

© Milliken Publishing Company  7  MP3378

Name_____

Least Common Denominator

CUT! CUT! CUT!

Roll the film. Find the least or lowest common denominator (LCD) and then write like fractions.

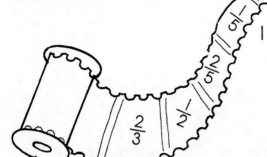

1. $\frac{2}{3}, \frac{1}{4}$

   LCD_____
   Like Fractions:
   _____
   _____

2. $\frac{1}{4}, \frac{1}{5}$

   LCD_____
   Like Fractions:
   _____
   _____

3. $\frac{2}{6}, \frac{3}{5}$

   LCD_____
   Like Fractions:
   _____
   _____

4. $\frac{5}{6}, \frac{1}{4}$

   LCD_____
   Like Fractions:
   _____
   _____

5. $\frac{2}{3}, \frac{3}{5}$

   LCD_____
   Like Fractions:
   _____
   _____

6. $\frac{1}{8}, \frac{2}{3}$

   LCD_____
   Like Fractions:
   _____
   _____

7. $\frac{2}{7}, \frac{1}{2}$

   LCD_____
   Like Fractions:
   _____
   _____

8. $\frac{1}{2}, \frac{5}{6}, \frac{2}{3}$

   LCD_____
   Like Fractions:
   _____
   _____

9. $\frac{1}{5}, \frac{1}{10}, \frac{2}{10}$

   LCD_____
   Like Fractions:
   _____
   _____

10. $\frac{4}{6}, \frac{1}{8}, \frac{2}{24}$

    LCD_____
    Like Fractions:
    _____
    _____

11. $\frac{1}{3}, \frac{1}{6}, \frac{2}{6}$

    LCD_____
    Like Fractions:
    _____
    _____

12. $\frac{3}{10}, \frac{2}{4}$

    LCD_____
    Like Fractions:
    _____
    _____

13. $\frac{5}{6}, \frac{1}{9}, \frac{3}{18}$

    LCD_____
    Like Fractions:
    _____
    _____

14. $\frac{1}{3}, \frac{3}{4}, \frac{2}{12}$

    LCD_____
    Like Fractions:
    _____
    _____

© Milliken Publishing Company

Name_____    Lowest Term Fractions

Write in lowest terms.

1. $\frac{2}{6} + \frac{3}{10}$  2. $\frac{3}{8} + \frac{1}{2}$  3. $\frac{3}{10} + \frac{1}{5}$  4. $\frac{2}{4} + \frac{3}{5}$  5. $\frac{2}{3} + \frac{4}{8}$

6. $\frac{7}{9} + \frac{2}{18}$  7. $\frac{4}{6} + \frac{3}{4}$  8. $\frac{2}{5} + \frac{3}{6}$  9. $\frac{4}{8} + \frac{1}{2}$  10. $\frac{4}{16} + \frac{2}{8}$

Subtract. Remember to write in lowest terms.

1. $\frac{3}{5} - \frac{1}{2}$  2. $\frac{6}{8} - \frac{1}{2}$  3. $\frac{6}{10} - \frac{1}{5}$  4. $\frac{2}{3} - \frac{1}{2}$  5. $\frac{3}{8} - \frac{1}{10}$

6. $\frac{5}{6} - \frac{2}{3}$  7. $\frac{5}{12} - \frac{2}{8}$  8. $\frac{4}{5} - \frac{2}{4}$  9. $\frac{2}{3} - \frac{1}{8}$  10. $\frac{4}{8} - \frac{1}{6}$

11. $\frac{6}{7} - \frac{2}{14}$  12. $\frac{3}{9} - \frac{2}{18}$  13. $\frac{3}{4} - \frac{1}{5}$  14. $\frac{7}{10} - \frac{1}{2}$  15. $\frac{2}{3} - \frac{1}{4}$

Addition answers correct (show as fraction) _____

Subtraction answers correct (show as fraction) _____

© Milliken Publishing Company

Name_____  Simplest Form

# FRACTURED FRACTIONS

Frieda fractured her favorite mirror. Reduce the fractions in each fragment.

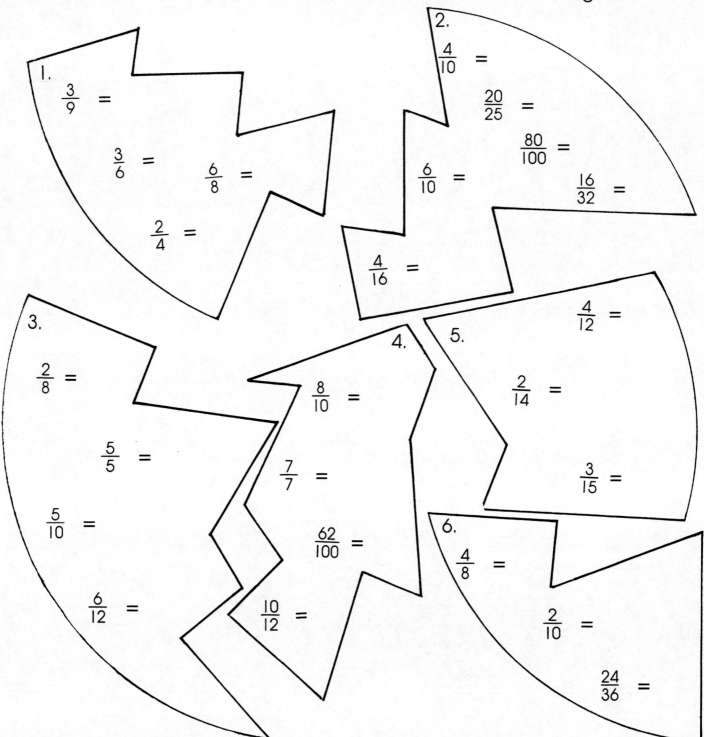

1. $\frac{3}{9} =$
   $\frac{3}{6} =$
   $\frac{6}{8} =$
   $\frac{2}{4} =$

2. $\frac{4}{10} =$
   $\frac{20}{25} =$
   $\frac{80}{100} =$
   $\frac{6}{10} =$
   $\frac{16}{32} =$
   $\frac{4}{16} =$

3. $\frac{2}{8} =$
   $\frac{5}{5} =$
   $\frac{5}{10} =$
   $\frac{6}{12} =$

4. $\frac{8}{10} =$
   $\frac{7}{7} =$
   $\frac{62}{100} =$
   $\frac{10}{12} =$

5. $\frac{4}{12} =$
   $\frac{2}{14} =$
   $\frac{3}{15} =$

6. $\frac{4}{8} =$
   $\frac{2}{10} =$
   $\frac{24}{36} =$

Write your score for each fragment. Write your scores as fractions.

1_____   2_____   3_____   4_____   5_____   6_____

Name_____          Reducing Fractions

# FRACTION FACTORY

**Reduce.**

- $\dfrac{18}{24} = \square$
- $\dfrac{9}{21} = \square$
- $\dfrac{10}{16} = \square$
- $\dfrac{9}{18} = \square$
- $\dfrac{6}{12} = \square$
- $\dfrac{21}{24} = \square$
- $\dfrac{9}{12} = \square$
- $\dfrac{10}{35} = \square$
- $\dfrac{7}{42} = \square$
- $\dfrac{25}{100} = \square$

**Reduce.**

- $\dfrac{4}{16} = \square$
- $\dfrac{8}{10} = \square$
- $\dfrac{15}{18} = \square$
- $\dfrac{6}{20} = \square$
- $\dfrac{3}{15} = \square$
- $\dfrac{18}{20} = \square$
- $\dfrac{12}{15} = \square$
- $\dfrac{4}{20} = \square$

**Match.**

| | |
|---|---|
| $\dfrac{16}{28}$ | $\dfrac{3}{7}$ |
| $\dfrac{6}{24}$ | $\dfrac{4}{7}$ |
| $\dfrac{15}{24}$ | $\dfrac{7}{8}$ |
| $\dfrac{18}{42}$ | $\dfrac{1}{2}$ |
| $\dfrac{21}{24}$ | $\dfrac{1}{4}$ |
| $\dfrac{50}{100}$ | $\dfrac{5}{8}$ |

© Milliken Publishing Company

Name_____

**FRACTION FLAVORS**

Proper, Improper, Mixed Number Fractions

Refresh your memory on three kinds of fractions.
Label (or shade) the ice cream cones below with the correct color.

$\frac{1}{2}, \frac{9}{12}, \frac{6}{8}, \frac{3}{5}$ (vanilla–white) Proper Fractions

$\frac{2}{2}, \frac{8}{4}, \frac{12}{5}, \frac{10}{6}$ (strawberry–red) Improper Fractions

$1\frac{1}{2}, 9\frac{3}{10}, 3\frac{4}{5}$ Mixed Numbers (chocolate–brown)

A.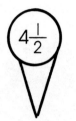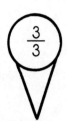

$4\frac{1}{2}$     $\frac{3}{3}$     $\frac{14}{17}$

B.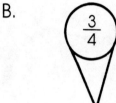

$\frac{3}{4}$     $6\frac{1}{8}$     $\frac{30}{10}$

C. 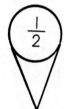    (and more)

$\frac{1}{2}$     $8\frac{3}{10}$     $\frac{16}{8}$     $8\frac{2}{3}$     $4\frac{1}{5}$     $\frac{2}{3}$

Write the following improper fractions as whole numbers or as mixed numbers. Write your answers on the cones.

D.

$\frac{17}{8}$     $\frac{14}{3}$     $\frac{11}{9}$     $\frac{25}{6}$     $\frac{4}{4}$     $\frac{18}{9}$     $\frac{22}{7}$     $\frac{9}{8}$

Now, rename these mixed numbers as improper fractions.

E.

$4\frac{1}{2}$     $5\frac{2}{3}$     $6\frac{3}{5}$     $2\frac{5}{8}$     $6\frac{1}{7}$     $3\frac{3}{10}$     $6\frac{4}{6}$     $1\frac{2}{5}$

© Milliken Publishing Company

MP3378

Name_____

Adding Fractions

# FRACTION FALLS

Add the water drop fractions.   Write your answers in the lowest terms.

$\frac{2}{3} + \frac{2}{3} =$

$\frac{3}{8} + \frac{7}{8} =$

$\frac{2}{9} + \frac{7}{9} =$

$\frac{4}{6} + \frac{5}{6} =$

$\frac{3}{8} + \frac{5}{8} =$

$\frac{5}{6} + \frac{4}{6} =$

$\frac{4}{7} + \frac{1}{7} =$

$\frac{7}{16} + \frac{4}{16} =$

$\frac{3}{10} + \frac{3}{10} =$

$\frac{3}{4} + \frac{3}{4} =$

$\frac{2}{3} + \frac{2}{3} =$

$\frac{4}{5} + \frac{4}{5} =$

$\frac{1}{8} + \frac{5}{8} =$

$\frac{3}{5} + \frac{1}{5} =$

$\frac{1}{2} + \frac{1}{2} =$

$\frac{6}{7} + \frac{2}{7} =$

$\frac{6}{10} + \frac{8}{10} =$

$\frac{5}{16} + \frac{5}{16} =$

$\frac{6}{12} + \frac{6}{12} =$

$\frac{8}{10} + \frac{1}{10} =$

$\frac{2}{4} + \frac{1}{4} =$

$\frac{3}{16} + \frac{3}{16} =$

$\frac{4}{6} + \frac{4}{6} =$

$\frac{7}{8} + \frac{7}{8} =$

$\frac{4}{7} + \frac{5}{7} =$

© Milliken Publishing Company          MP3378

Name_____    Subtracting Fractions

## SUBTRACTING IN SPACE

Solve these subtraction problems.

1. $\dfrac{5}{8} - \dfrac{3}{8} =$

2. $\dfrac{7}{10} - \dfrac{6}{10} =$

3. $\dfrac{4}{5} - \dfrac{1}{5} =$    4. $\dfrac{9}{8} - \dfrac{4}{8} =$    5. $\dfrac{4}{6} - \dfrac{3}{6} =$

6. $\dfrac{9}{18} - \dfrac{7}{18} =$    7. $\dfrac{9}{16} - \dfrac{5}{16} =$    8. $\dfrac{4}{8} - \dfrac{4}{8} =$

9. $\dfrac{7}{9} - \dfrac{5}{9} =$    10. $\dfrac{7}{10} - \dfrac{4}{10} =$    11. $\dfrac{4}{6} - \dfrac{2}{6} =$

12. $\dfrac{7}{10} - \dfrac{6}{10} =$    13. $\dfrac{9}{10} - \dfrac{4}{10} =$    14. $\dfrac{5}{8} - \dfrac{3}{8} =$

15. $\dfrac{3}{7} - \dfrac{2}{7} =$    16. $\dfrac{5}{6} - \dfrac{3}{6} =$    17. $\dfrac{3}{5} - \dfrac{1}{5} =$

18. $\dfrac{7}{8} - \dfrac{3}{8} =$    19. $\dfrac{2}{3} - \dfrac{1}{3} =$

20. $\dfrac{7}{10} - \dfrac{3}{10} =$

© Milliken Publishing Company    MP3378

Name_____   Adding Unlike Fractions

## ADDING FUNKY FRACTIONS

Add the following unlike fractions.

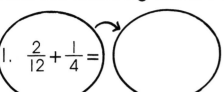
$\dfrac{5}{8} + \dfrac{3}{4} =$   $\dfrac{5}{8} + \dfrac{6}{8} = \dfrac{11}{8}$ or $1\dfrac{3}{8}$

1. $\dfrac{2}{12} + \dfrac{1}{4} =$

2. $\dfrac{5}{10} + \dfrac{2}{3} =$

3. $\dfrac{1}{4} + \dfrac{10}{16} =$

4. $\dfrac{6}{10} + \dfrac{1}{2} =$

5. $\dfrac{4}{8} + \dfrac{6}{10} =$

6. $\dfrac{14}{60} + \dfrac{1}{2} =$

7. $\dfrac{1}{5} + \dfrac{2}{10} =$

8. $\dfrac{1}{4} + \dfrac{12}{24} =$

9. $\dfrac{1}{6} + \dfrac{2}{18} =$

10. $\dfrac{13}{5} + \dfrac{6}{15} =$

11. $\dfrac{5}{25} + \dfrac{7}{5} =$

12. $\dfrac{1}{12} + \dfrac{1}{4} =$

13. $\dfrac{10}{35} + \dfrac{2}{7} =$

14. $\dfrac{20}{40} + \dfrac{1}{2} =$

15. $\dfrac{16}{20} + \dfrac{2}{5} =$

Total Awesome Score

Name_____

Subtracting and
Renaming Fractions

**Subtract and rename these fractions.**

A.  $82$      $10\frac{3}{6}$      $11\frac{2}{10}$      $14\frac{2}{7}$
   $-\frac{6}{8}$   $-2\frac{4}{6}$   $-3\frac{4}{10}$   $-7\frac{6}{7}$

B.  $62\frac{5}{12}$   $14\frac{2}{7}$   $5\frac{3}{10}$   $9\frac{3}{8}$
   $-40\frac{7}{12}$   $-6\frac{6}{7}$   $-2\frac{7}{10}$   $-4\frac{7}{8}$

C.  $10\frac{1}{5}$   $6\frac{1}{8}$   $18\frac{2}{16}$   $22\frac{2}{14}$
   $-5\frac{3}{5}$   $-3\frac{4}{8}$   $-17\frac{5}{16}$   $-20\frac{3}{14}$

D.  $85\frac{1}{6}$   $2\frac{8}{10}$   $6\frac{1}{5}$   $4\frac{1}{8}$
   $-1\frac{9}{6}$   $-1\frac{9}{10}$   $-2\frac{4}{5}$   $-2\frac{3}{8}$

E.  $6\frac{1}{9}$   $18$   $4\frac{5}{10}$   $6$
   $-2\frac{6}{9}$   $-\frac{2}{8}$   $-2\frac{9}{10}$   $-3\frac{5}{7}$

F.  $34\frac{7}{10}$   $20\frac{8}{12}$   $5\frac{6}{10}$   $39\frac{3}{8}$
   $-30\frac{9}{10}$   $-10\frac{9}{12}$   $-1\frac{3}{10}$   $-29\frac{5}{8}$

© Milliken Publishing Company

MP3378

Name_____

Dividing Mixed, Whole and Fractional Numbers

Write each quotient in simplest form.

A. $6 \div 1\frac{2}{5} =$ $\qquad$ $16 \div 4\frac{1}{8} =$ $\qquad$ $20 \div 2\frac{1}{2} =$

B. $42 \div 6\frac{1}{7} =$ $\qquad$ $1\frac{2}{3} \div 4 =$ $\qquad$ $4\frac{1}{2} \div 5 =$

C. $1\frac{2}{5} \div 10 =$ $\qquad$ $1\frac{7}{8} \div 12 =$ $\qquad$ $1\frac{1}{4} \div 2 =$

D. $2\frac{2}{10} \div \frac{3}{4} =$ $\qquad$ $3\frac{1}{6} \div \frac{1}{3} =$ $\qquad$ $4\frac{1}{2} \div \frac{1}{4} =$

E. $\frac{1}{5} \div 2\frac{1}{2} =$ $\qquad$ $\frac{2}{8} \div 6\frac{1}{4} =$ $\qquad$ $3\frac{3}{5} \div \frac{2}{10} =$

F. $1\frac{1}{7} \div \frac{5}{6} =$ $\qquad$ $2\frac{1}{8} \div \frac{2}{5} =$ $\qquad$ $\frac{6}{7} \div 4\frac{1}{3} =$

G. $1\frac{1}{6} \div 4\frac{1}{3} =$ $\qquad$ $3\frac{1}{10} \div 2\frac{1}{5} =$ $\qquad$ $3\frac{2}{5} \div 1\frac{1}{2} =$

H. $3\frac{1}{2} \div 1\frac{1}{8} =$ $\qquad$ $2\frac{2}{3} \div 1\frac{1}{12} =$ $\qquad$ $2\frac{4}{9} \div 1\frac{1}{9} =$

I. $4\frac{5}{6} \div 1\frac{1}{6} =$ $\qquad$ $4\frac{5}{10} \div 2\frac{2}{5} =$ $\qquad$ $1\frac{2}{5} \div 3\frac{6}{10} =$

© Milliken Publishing Company

Name_____

Multiplying Fractions

# FRACTION FLAG

Find each product. Reduce your answers to lowest terms. Circle your correct answers with a blue pencil.

1. $\dfrac{6}{9} \times \dfrac{3}{7} =$

2. $\dfrac{3}{5} \times \dfrac{2}{8} =$

3. $\dfrac{1}{3} \times \dfrac{4}{5} =$

4. $\dfrac{2}{12} \times \dfrac{3}{5} =$

5. $\dfrac{13}{15} \times \dfrac{2}{3} =$

6. $\dfrac{9}{12} \times \dfrac{1}{2} =$

7. $\dfrac{3}{4} \times \dfrac{3}{7} =$

8. $\dfrac{2}{10} \times \dfrac{4}{14} =$

9. $\dfrac{1}{12} \times \dfrac{4}{5} =$

10. $\dfrac{3}{8} \times \dfrac{4}{9} =$

11. $\dfrac{1}{2} \times \dfrac{10}{12} =$

12. $\dfrac{4}{10} \times \dfrac{5}{8} =$

13. $\dfrac{2}{5} \times \dfrac{10}{16} =$

14. $\dfrac{1}{4} \times \dfrac{8}{12} =$

15. $\dfrac{1}{3} \times \dfrac{9}{15} =$

16. $\dfrac{1}{6} \times \dfrac{12}{24} =$

17. $\dfrac{1}{8} \times \dfrac{16}{20} =$

18. $\dfrac{7}{10} \times \dfrac{5}{8} =$

19. $\dfrac{3}{10} \times \dfrac{5}{7} =$

20. $\dfrac{4}{20} \times \dfrac{5}{8} \times \dfrac{8}{9} =$

20 blue circles = Fantastic
15 blue circles = Fairly wonderful
10 blue circles = Fair
 5 blue circles = Fortune has not smiled on you.

Name_____

Dividing Fractions

## A FORTUNE IN FRACTIONS

Find the quotients. Reduce, if possible.

1. $\frac{10}{5} \div 2 =$
2. $\frac{12}{19} \div 6 =$
3. $4 \div \frac{7}{9} =$
4. $8 \div \frac{1}{2} =$
5. $\frac{6}{2} \div 2 =$
6. $\frac{8}{9} \div 9 =$
7. $10 \div \frac{2}{5} =$
8. $\frac{14}{7} \div 7 =$
9. $\frac{10}{20} \div 5 =$
10. $\frac{1}{2} \div 3 =$

11. $5 \div \frac{4}{9} =$
12. $\frac{2}{3} \div 6 =$
13. $\frac{5}{8} \div 16 =$
14. $14 \div \frac{2}{7} =$
15. $15 \div \frac{3}{5} =$
16. $16 \div \frac{2}{8} =$
17. $\frac{16}{17} \div 4 =$
18. $\frac{4}{7} \div 2 =$
19. $\frac{3}{8} \div 6 =$
20. $\frac{8}{9} \div 2 =$

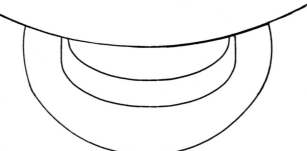

$\overline{20}$

Is your future bright with the right answers? Show your fraction of correct answers.

Name _____       Adding Fractions

\# correct

28. $50\frac{1}{3}$
    $+95\frac{5}{8}$

\# correct _____

A flight of stairs
From easy to hard—
Can you reach the top
With no errors?

26. $47\frac{4}{9}$    27. $6\frac{1}{2}$
    $+32\frac{1}{6}$      $+8\frac{5}{12}$

\# correct _____

23. $62\frac{1}{2}$    24. $5\frac{2}{3}$    25.
    $40\frac{2}{3}$        $4\frac{5}{8}$         $20\frac{16}{20}$
    $+89\frac{2}{5}$      $+3\frac{5}{6}$       $+89\frac{7}{10}$

\# correct _____

19. $8\frac{4}{9}$    20.               21.              22.
    $6\frac{8}{9}$        $3\frac{4}{5}$       $6\frac{1}{3}$      $14\frac{1}{4}$
    $+12\frac{2}{9}$    $+1\frac{1}{2}$    $+8\frac{5}{12}$  $+9\frac{7}{8}$

\# correct _____

                              17. $18\frac{5}{8}$   18. $9\frac{8}{10}$
14. $49\frac{3}{9}$  15. $56\frac{4}{10}$  16. $72\frac{6}{8}$   $10\frac{7}{8}$       $12\frac{7}{10}$
    $+62\frac{8}{9}$      $+12\frac{9}{10}$      $+11\frac{5}{8}$      $+2\frac{4}{8}$      $+3\frac{4}{10}$

\# correct _____

8. $2\frac{1}{8}$  9. $3\frac{3}{9}$  10. $4\frac{2}{5}$  11. $8\frac{11}{12}$  12. $4\frac{6}{8}$  13. $24\frac{6}{10}$
   $+6\frac{5}{8}$    $+2\frac{3}{9}$     $+3\frac{3}{5}$       $+10\frac{5}{12}$     $+18\frac{3}{8}$     $+14\frac{5}{10}$

\# correct _____

1. $2\frac{1}{4}$  2. $1\frac{1}{8}$  3. $6\frac{1}{2}$  4. $4\frac{3}{5}$  5. $2\frac{2}{7}$  6. $6\frac{1}{7}$  7. $5\frac{2}{10}$
   $+2\frac{1}{4}$    $+3\frac{2}{8}$    $+6\frac{1}{2}$    $+4\frac{1}{5}$    $+1\frac{1}{7}$    $+4\frac{1}{7}$    $+5\frac{3}{10}$

\# correct _____

Name_____          Subtracting Fractions

## FRACTIONS IN FLIGHT

Find each difference.

A.  $7\frac{7}{8}$          $9\frac{9}{10}$          $5\frac{1}{2}$          $25\frac{8}{9}$
   $-4\frac{6}{8}$       $-6\frac{5}{10}$       $-1$              $-14\frac{7}{9}$

B.  $15\frac{1}{8}$         $14$              $84\frac{8}{20}$        $5\frac{3}{10}$
   $-12\frac{7}{8}$      $-8\frac{3}{16}$      $-44\frac{18}{20}$      $-1\frac{8}{10}$

C.  $7\frac{7}{8}$          $16\frac{8}{10}$        $42\frac{3}{5}$         $29\frac{4}{5}$
   $-4\frac{3}{4}$       $-12\frac{2}{20}$     $-31\frac{1}{15}$      $-21\frac{1}{3}$

D.  $26\frac{7}{9}$         $10\frac{6}{9}$         $94\frac{3}{24}$        $17\frac{3}{4}$
   $-12\frac{3}{4}$      $-1\frac{1}{3}$       $-16\frac{1}{8}$       $-8\frac{3}{8}$

E.  $29\frac{5}{6}$         $15\frac{1}{2}$         $8\frac{1}{2}$          $47\frac{2}{3}$
   $-3\frac{1}{2}$       $-9\frac{3}{8}$       $-3\frac{5}{8}$        $-13\frac{3}{5}$

F.  $63\frac{2}{5}$         $37\frac{1}{2}$         $82\frac{4}{9}$         $75\frac{3}{10}$
   $-49\frac{1}{6}$      $-4\frac{1}{7}$       $-17\frac{1}{5}$       $-26\frac{2}{3}$

© Milliken Publishing Company          MP3378

Name_____   Fractions as Decimals

Write these fractions as decimals.

A.  $4\frac{3}{10}$ = _____   $2\frac{8}{10}$ = _____   $24\frac{7}{100}$ = _____   $\frac{8}{1000}$ = _____

B.  $16\frac{32}{100}$ = _____   $\frac{285}{1000}$ = _____   $\frac{360}{10,000}$ = _____   $\frac{5}{10}$ = _____

Change these decimals to fractions or mixed numbers.

C.  0.70 = _____   .04 = _____   0.311 = _____   14.005 = _____

D.  8.2 = _____   24.8 = _____   0.0004 = _____   3.15 = _____

Add these fractions. Express as decimals.

E.  $\frac{3}{10} + \frac{4}{100} + \frac{6}{1000}$ = _____   $2 + \frac{6}{100} + \frac{2}{10} + \frac{8}{10}$ = _____

F.  $40 + \frac{2}{10} + \frac{8}{100} + \frac{20}{100}$ = _____   $\frac{6}{100} + \frac{6}{10,000}$ = _____

G.  $\frac{14}{100} + \frac{3}{10} + \frac{5}{10} + \frac{8}{10}$ = _____   $\frac{50}{100} + \frac{10}{100} + \frac{80}{10,000}$ = _____

Subtract these fractions. Express as decimals.

H.  $\frac{8}{10} - \frac{6}{10}$ = _____   $\frac{50}{100} - \frac{27}{100}$ = _____   $\frac{80}{10,000} - \frac{40}{10,000}$ = _____

I.  $8\frac{16}{100} - 4\frac{4}{100}$ = _____   $82\frac{8}{10} - 52\frac{6}{10}$ = _____   $\frac{30}{10,000} - \frac{6}{10,000}$ = _____

Name_____       Multiplying Decimal Fractions

Multiply these decimal fractions.

A.  $\begin{array}{r}1.2\\ \times\ \ .3\\ \hline\end{array}$   $\begin{array}{r}6.8\\ \times\ \ .2\\ \hline\end{array}$   $\begin{array}{r}24.1\\ \times\ .05\\ \hline\end{array}$   $\begin{array}{r}6.05\\ \times\ .02\\ \hline\end{array}$   $\begin{array}{r}16.1\\ \times\ \ .4\\ \hline\end{array}$

B.  $\begin{array}{r}10.8\\ \times\ \ .4\\ \hline\end{array}$   $\begin{array}{r}15.6\\ \times\ .35\\ \hline\end{array}$   $\begin{array}{r}20.2\\ \times\ \ .6\\ \hline\end{array}$   $\begin{array}{r}85.9\\ \times\ .52\\ \hline\end{array}$   $\begin{array}{r}46.9\\ \times\ .08\\ \hline\end{array}$

Multiply. Express as a decimal.

C.  $\frac{2}{10} \times \frac{3}{100}$   $\frac{8}{10} \times \frac{4}{10}$   $\frac{16}{100} \times \frac{6}{10}$   $\frac{5}{100} \times \frac{6}{100}$   $\frac{3}{10} \times \frac{3}{10}$

D.  $\frac{9}{10} \times \frac{2}{10}$   $\frac{20}{100} \times \frac{2}{100}$   $\frac{1}{10} \times \frac{40}{100}$   $\frac{14}{100} \times \frac{20}{100}$   $\frac{2}{10} \times \frac{5}{10}$

E.  $\frac{4}{100} \times \frac{8}{100}$   $\frac{42}{100} \times \frac{2}{10}$   $\frac{6}{100} \times \frac{4}{10}$   $\frac{20}{1000} \times \frac{2}{1000}$   $\frac{4}{10} \times \frac{6}{100}$

© Milliken Publishing Company                                              MP3378

Name_____    Dividing Decimal Fractions

Divide these decimal fractions. Express as a decimal.

A.  $14\frac{49}{100} \div 2\frac{3}{10} =$ _____     $4\frac{62}{100} \div 2\frac{31}{100} =$ _____     $88\frac{8}{10} \div 11\frac{1}{10} =$ _____

B.  $6\frac{80}{100} \div 3\frac{40}{100} =$ _____     $52\frac{9}{10} \div 4\frac{6}{10} =$ _____     $15\frac{6}{10} \div 5\frac{2}{10} =$ _____

C.  $50\frac{20}{100} \div 25\frac{1}{10} =$ _____     $8\frac{4}{10} \div 4\frac{2}{10} =$ _____     $7\frac{854}{1000} \div 7\frac{7}{100} =$ _____

Write each quotient in simplest form.

D.  $3\frac{3}{5} \div 2\frac{2}{5} =$ _____     $2\frac{1}{2} \div 3\frac{2}{3} =$ _____     $7\frac{1}{2} \div 3\frac{3}{4} =$ _____

E.  $2\frac{4}{5} \div 5\frac{3}{4} =$ _____     $5\frac{5}{6} \div 4\frac{2}{3} =$ _____     $2\frac{7}{9} \div 2\frac{3}{4} =$ _____

F.  $1\frac{1}{14} \div 2\frac{1}{12} =$ _____     $3\frac{5}{8} \div 14\frac{1}{2} =$ _____     $2\frac{1}{10} \div 4\frac{1}{2} =$ _____

G.  $2\frac{1}{4} \div 1\frac{5}{6} =$ _____     $3\frac{3}{4} \div 12\frac{1}{2} =$ _____     $7\frac{1}{8} \div 1\frac{1}{5} =$ _____

Name_____    Tenths and Hundredths

## PARTS OF A WHOLE

Write the decimal for each shaded part below.

_____

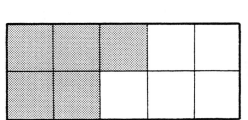
_____

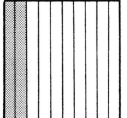
_____

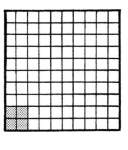
_____

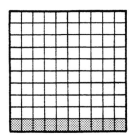
_____

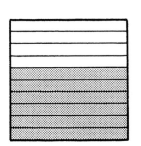
_____

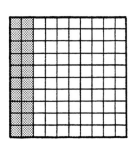
_____

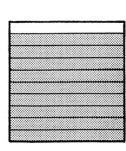
_____

Write as decimals.

Three pennies equal what part of a dime?

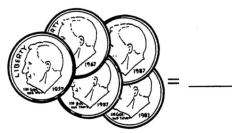

Five dimes equal what part of a dollar?

Write as decimals.
1. One-half a mile equals _____ of a mile.
2. Three dimes equal _____ of a dollar?
3. Thirty pennies equal _____ of a dollar.
4. The letters **H** and **U** equal _____ of the word **H U N D R E D T H S.**
5. Eight pennies equal _____ of a dime.
6. The toes on your right foot equal _____ of all your toes.
7. One millimeter equals _____ of a centimeter.
8. Sixty centimeters equal _____ of a meter.
9. Three quarters equal _____ of a dollar.

Name_____   Writing Decimals

# E - Z QUIZ

Q. How many zeros are there in ten?
A. _____
Q. How many places are there to the right of the zero in tenths?
A. _____
Q. How many zeros are there in hundredths?
A. _____
Q. How many places are there to the right of the zero in hundredths?
A. _____

Write these decimals as you read them from this list.

eight tenths _____    seventeen hundredths _____
one and eight tenths _____    thirty-seven and four tenths _____
two and four tenths _____    ninety-one hundredths _____
six tenths _____    one and one tenth _____
eight and twenty-nine hundredths ____    seven tenths _____
seven and eight tenths _____    two and eight tenths _____
two and four hundredths _____    five tenths _____
five and thirty-eight hundredths _____    one and ninety-two hundredths _____
nine tenths _____    nine and eighty hundredths _____
eighty-five hundredths _____    one and twenty-three hundredths ____

Write these decimals as words.

7.8 _____    1.38 _____
1.05 _____    3.9 _____
21.6 _____    8.08 _____
15.0 _____    .68 _____
5.6 _____    .10 _____
1.60 _____    7.9 _____
.07 _____    .4 _____
6.6 _____    .50 _____
.2 _____    5. _____
142.42 _____    .5 _____

© Milliken Publishing Company    26    MP3378

Name_____

Comparing Decimals

### NUMBERS
There are 3 of us in two,
Five of us in seven,
Four of us in nine,
And, six of us in eleven.

Write < or > to compare these decimals.
(Hint: compare digits, left to right.)

A.  0.1 ___ 1.0      1.5 ___ 1.8      3.06 ___ 3.60
B.  0.6 ___ 0.9      4.1 ___ 4.3      4.71 ___ 47.1
C.  0.5 ___ 0.7      6.7 ___ 7.6      6.52 ___ 6.51
D.  0.15 ___ 0.18    0.09 ___ 0.08    7.7 ___ 7.3

Find the answer to this riddle by writing the decimals in order, from least to greatest.

| S | T | E | L | R | E | T |
|---|---|---|---|---|---|---|
| 26.4 | 22.8 | 2.73 | 2.61 | 26.0 | 25.1 | 3.70 |

Write the decimals in order from the least to the greatest.

E. 6.2, 4.3, 5.7 _____

F. 5.12, 51.2, 5.1 _____

G. 7.34, 7.4, 7.36 _____

H. 4.2, 42, 4.21 _____

I. Each scoop of ice cream is smaller than the one below it. Number the scoops in order (from top to bottom) from least to greatest.

3.41, 3.62, 3.21, 3.33, 3.31

Name_____                                    Adding Decimals

SUM IT UP!

Write out these problems. Remember to keep the decimal points in a line.

$2.4 + 7.8 =$ _____
```
  2.4
+ 7.8
-----
 10.2
```

| A. | | | |
|---|---|---|---|
| $3.6 + 1.5 =$ _____ | $1.06 + 4.60 =$ _____ | $41.60 + 7.8 =$ _____ |

| B. | | | |
|---|---|---|---|
| $31.4 + 7.61 =$ _____ | $9.6 + 0.21 =$ _____ | $34.73 + 5.46 =$ _____ | $5.6 + 7.0 =$ _____ |

Write the sum.

C.
```
   7.4        6.79       42.1        7.60       15.37
   4.4        2.8        65.42       7.94        3.08
   8.64       8.06        5.90      97.08        7.04
+  5.73     + 7.8      + 16.56    + 64.70     + 16.35
```

D.
```
  52.36       7.40       95.60      13.53       75.6
   7.50       5.35        7.3        7.04        0.9
   4.56       7.90       78.43       1.16        6.2
+  0.70     + 4.5      +  5.64    +  8.43     + 70.53
```

E. Put on your hiking boots! From base camp to the clouds you hiked 2.1 miles, through the clouds another .06 miles, came to snow by hiking .7 more miles, and reached the summit by hiking the last .87 miles. How far did you hike?

_____

Name_____  Subtracting Decimals

## COMING DOWN

Find the difference. Remember to line up the decimal points!

A.  6.1 − 3.9 = ____    9.26 − 3.74 = ____    8.31 − 2.07 = ____

B.  6.04 − 5.89 = ____    51.49 − 29.25 = ____    8.00 − 4.62 = ____

60.14 − 37.28 = ____

Find the difference.

| | | | | |
|---|---|---|---|---|
| C.   3.2 <br> − 2.2 | 29.38 <br> − 24.69 | 70.59 <br> − 12.62 | 63.27 <br> −  .49 | 23.71 <br> −  .08 |
| D.   59.2 <br> − 21.3 | 7.81 <br> − 3.13 | 8.6 <br> − 2.1 | 6.87 <br> − 3.51 | 28.20 <br> − 27.54 |
| E.   6.28 <br> − 2.46 | 8.71 <br> − 8.34 | 1.21 <br> − 0.23 | 7.61 <br> − 2.82 | 6.4 <br> − 2.2 |

Find the missing number to balance the weights.

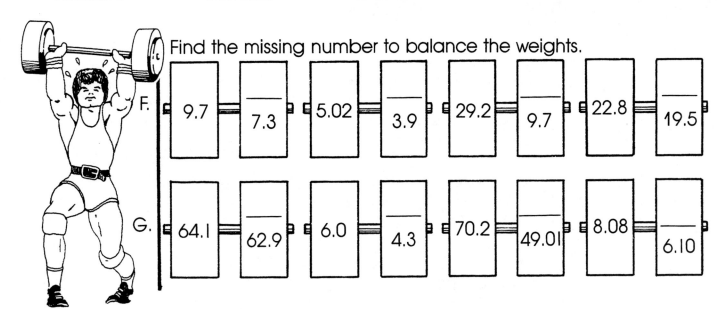

F.  9.7 | 7.3 | 5.02 | 3.9 | 29.2 | 9.7 | 22.8 | 19.5

G.  64.1 | 62.9 | 6.0 | 4.3 | 70.2 | 49.01 | 8.08 | 6.10

Name_____     Decimals and Money

## MAKING SENSE OUT OF DECIMALS

$1.00 = 1 and 0 tenths, 0 hundredths

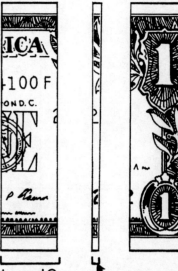

50¢ or .50 = 50 hundredths of a dollar

25¢ or .25 = 25 hundredths of a dollar

10¢ or .10 = 10 hundredths of a dollar

1¢ or .01 = 1 hundredths of a dollar

Find the sum or difference. (Notice the decimal points are in a straight line.)

A.
```
  $2.50        $4.80        $1.41        $ .50        $4.98
   .50          .55         1.50         1.76          .42
 + .25        + .48        + .75        +2.60        +1.98
```

B.
```
  $2.89        $4.71        $5.78        $3.70        $6.20
 - .47        -2.37        -3.87        - .50        -1.76
```

Write out these problems. Find the sum or difference.

C. ADD seven dollars and fifty cents,
   three dollars and forty-two cents,
   two dollars and twenty-nine cents.  _____

D. SUBTRACT four dollars and twenty-nine cents from
   thirty-three dollars and forty-nine cents.  _____

E. ADD forty-eight dollars and sixty cents,
   twenty-three dollars and forty-five cents,
   three dollars and eight-five cents.  _____

F. SUBTRACT two dollars and twenty-nine cents from
   five dollars and fifteen cents.  _____

Name_____

Place Value

# TENTHS, HUNDREDTHS, THOUSANDTHS

Write as decimals. Remember to use a decimal point for the word "and".

A. 42 hundredths _____
B. 3 tenths _____
C. 32 hundredths _____
D. 561 thousandths _____
E. 423 thousandths _____
F. 42 thousandths _____
G. 2 thousandths _____

2 and 42 thousandths _____
614 thousandths _____
6 and 41 thousandths _____
40 and 2 thousandths _____
7 and 69 thousandths _____
4 and 476 thousandths _____
87 and 87 thousandths _____

Write the value of the underlined digits.

H. 1.<u>7</u>62 _____
I. 7.82<u>0</u> _____
J. 4.32<u>1</u> _____
K. 6.8<u>7</u> _____

0.82<u>5</u> _____
0.0<u>9</u>4 _____
0.93<u>2</u> _____
1.06<u>8</u> _____

What is this page all about?
Write the letter in front of the correct answer.

| | | | |
|---|---|---|---|
| E | Four is in the tenths place. | ___ | 9.724 |
| I | Two is in the thousandths place. | ___ | 972.40 |
| A | Five is in the tenths place. | ___ | 97.24 |
| S | Eight is in the hundredths place. (The number is more than one.) | ___ | 4.612 |
| C | Four is in the hundredths place. | ___ | 7.677 |
| D | Four is in the thousandths place. | ___ | 85.57 |
| L | Eight is in the hundredths place. (The number is less than one.) | ___ | .689 |
| M | Seven is in the thousandths place. | ___ | 4.888 |

© Milliken Publishing Company

MP3378

Name_____    Relating Decimals and Fractions

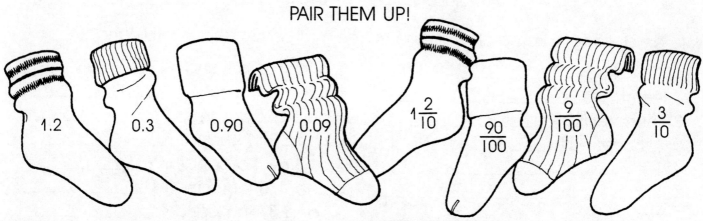

Write as decimals.

A. $\frac{8}{10}$ =     $\frac{4}{100}$ =

B. $\frac{8}{100}$ =    $\frac{4}{10}$ =

C. $\frac{80}{100}$ =   $\frac{40}{100}$ =

Write as fractions.

D. 0.60 =     0.3 =

E. 0.06 =     0.30 =

F. 0.6 =      0.03 =

Write as decimals.

G. $\frac{45}{100}$ =   $\frac{11}{100}$ =   $7\frac{30}{100}$ =   $12\frac{10}{100}$ =

H. $\frac{60}{100}$ =   $\frac{15}{100}$ =   $5\frac{60}{100}$ =   $28\frac{40}{100}$ =

I. $\frac{5}{100}$ =    $\frac{13}{100}$ =   $1\frac{40}{100}$ =   $3\frac{4}{100}$ =

J. $\frac{29}{100}$ =   $\frac{20}{100}$ =   $8\frac{4}{100}$ =    $9\frac{3}{100}$ =

K. $\frac{13}{100}$ =   $\frac{1}{100}$ =    $6\frac{5}{100}$ =    $9\frac{30}{100}$ =

Write as fractions or mixed numbers in simplest form.

L. .2      .75     .9      .05

M. .25     .3      .12     .35

N. 6.25    2.5     6.2     3.25

O. .37     .62     7.7     4.12

P. .28     5.75    .45     3.6

Name_____          Thousandths

LOOK WHAT'S AHEAD!

Addition and subtraction. Watch the signs.

A.  26.79      16.128     17.306
    14.320      9.89       6.412
   + 4.351    + 4.362    + 8.129

B.  .2014     5.201     42.176     5.7631     8.319
   −.0462    +4.202    +14.02    − 4.392    − .723

C.  2.2804    1.002    5.124    9.0601    3.726
   +.3453   − .08    − .619   − 8.2092    5.647
                                           7.303
                                            .092
                                           6.423
                                          + .99

D.  6.701    6.7293    6.0051    637.029
     .481   + .3624   − 3.016   + 421.704
    3.410
    2.605
     .048
   + .972

E.  .3627    4.318    .352     1.973
   −.0438   −2.605   −.2945   − 1.782

© Milliken Publishing Company    33    MP3378

Name_____

Estimating Sums and Differences

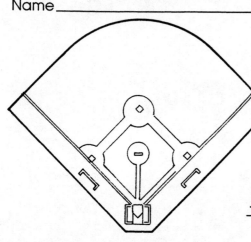

IT'S IN THE BALLPARK.

Ballpark figures can be used to estimate prices when you are shopping.

$45.10 = $45
+16.80 = 17
$61.90   $62

If the digit to the right of the decimal is 5 or greater, round the whole number up. If it is less than 5, round down.

The estimate of $62 is close enough. Round to the nearest dollar and estimate the sums. Then check the problems by adding the exact amounts and comparing your answers with your estimates.

A.  $18.60 = 19        $34.70 =         $9.10 =          $24.70 =
   +28.30 = 28        + 7.20 = __     +8.89 = __       + 8.30 = __

B.  $28.23 =          $47.19 =         $214.72 =        $18.97 =
     .43  =            1.72  =          972.16 =          7.60 =
    3.93  =            6.35  =             .84 =          9.80 =
   +4.57  = __        +3.95  = __       +7.49 = __       +5.82 = __

Round and estimate the differences. Check your estimate with the actual product.

C.  $72.36 =          $58.21 =         $24.62 =         $36.52 =
   - 9.26 = __       - 5.94 = __      - 7.23 = __      - 7.44 = __

At the Cash Register: Just write either yes or no and your estimate.
   Will $5 be enough to purchase these items?
   D.  One pair of socks $2.75, long shoestrings $1.69, comb $.79 _____

   Will $10 cover the cost of these items?                          _____
   E.  Two theater tickets $2.75 each, 2 hamburgers, $1.30 each, 2 medium sodas $.69 each _____

_____

Name_____

Multiplying Decimals By Whole Numbers

Multiply. Remember the decimal and the product will have the same number of decimal places.

| | | | | | |
|---|---|---|---|---|---|
| A. | 0.5 × 6 | 11.3 × 5 | 3.9 × 6 | 12.4 × 2 | 54.2 × 7 |
| B. | 7.3 × 6 | 16.4 × 7 | 8.4 × 8 | 4.01 × 6 | 7.343 × 3 |
| C. | 7.42 × 24 | 8.93 × 14 | .306 × 29 | 2.58 × 56 | .950 × 68 |
| D. | .218 × 80 | 8.4 × 16 | .750 × 35 | .035 × 18 | 0.18 × 24 |
| E. | 18.67 × 14 | 0.37 × 23 | 8.88 × 65 | 36.5 × 41 | 3.39 × 93 |

Calculator Quiz: 1334.5 × 4 = (Turn your calculator upside down to find the answers.)

Be careful of these if you're in a flower garden on a warm summer day. Problem answer _____ upside down _____

Name_____                    Estimating Products

## IS YOUR ANSWER REASONABLE?

A ballpark figure is useful to check multiplication.

    Find the first number in factor that is not a zero.      7.361 = 7

    Round each factor to that number.
    Multiply the two factors.
          ×4.8 = ×5
          58888    35

    Compare the estimated product with the actual product.
          29444
          35.3328

Estimate the product. Check your estimate with the actual product.

A.   .86 = .9          5.73 =          4.6 =
    ×.023 = .02      ×.062 =       ×3.2 =
    .01978  .018

compare

B.   .82 =          2.8 =          .72 =          5.4 =
    ×4.5 =        ×3.9 =       ×.48 =       ×.17 =

C.   .89 =          8.04 =         10.81 =       2.76 =
    ×.077 =       ×.06 =       ×2.04 =     ×.067 =

D.   4.85 =        12.02 =       .049 =        44.29 =
    ×.34 =        ×.042 =      ×.37 =       ×2.06 =

© Milliken Publishing Company

Name_____

Dividing Decimals

SMOOTH SAILING

Dividing decimals by a whole number? Remember to place the quotient decimal directly above the dividend decimal.

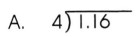

Write each quotient.

A.  4)1.16      6)54.6     3)1.2      4)1.6      9)87.3

B.  7)9.8       2)1.06     5)52.0     8)5.68     6)18.78

C.  34)8.84     16)9.6     21)7.728   38)7.6     32).768

D.  13)3.926    27)9.72    63)50.4    36)8.64    15)61.5

E.  82)33.292   16)5.232   43)34.486  15)3.750   36)84.24

The rainfall in Sailorville for one year was 78.24 inches. What was the average rainfall per month? _____

The distance across the lake at Sailorville is 115.71 km. How long will it take to cross the lake in a motorboat, traveling 29 km per hour? _____

© Milliken Publishing Company       37       MP3378

Name_____     Comparing Decimals

## <, >, =

Write <, >, or = and compare the decimals.
Hint: Mentally line up the decimal points and compare digits, left to right.

A.  0.164 ___ 0.621        3.7 ___ .57

B.  41.08 ___ 41.80        .012 ___ .013

C. 0420 ___ .03     7.10 ___ 7.1      77.00 ___ 77.03     7.80 ___ 7.8000

D. .002 ___ .003    4.1 ___ 4.01      1.42 ___ 1.421      .0894 ___ .09

E. 2.06 ___ 2.0067  .395 ___ .394     4.2000 ___ 4.2      .006 ___ .060

What animals don't come in pairs?
Arrange the decimals and their letters in order, least to greatest, to solve the riddle.

(F)  O     M     R     W        (G)  R     A     E     S
    .87   8.7   8.07  .087           .61   .16   1.06  .08

(H)  P      I      A      N      (I)  E     S     P     L
    8.456  5.678  7.875  6.784        .424  .427  .402  .404

Letters:   (F) ___ ___ ___ ___        (G) ___ ___ ___ ___

Decimals:  (F) ___ ___ ___ ___        (G) ___ ___ ___ ___

Letters:   (H) ___ ___ ___ ___        (I) ___ ___ ___ ___ !

Decimals:  (H) ___ ___ ___ ___        (I) ___ ___ ___ ___

© Milliken Publishing Company                                    MP3378

Name _____   Fractions to Decimals

## CALCULATOR CHALLENGE

Using a calculator? Change these fractions to decimals. Write these fractions as decimals, to the nearest hundredth.

$$\frac{1}{2} = 2\overline{)1.00} \qquad \frac{1}{2} = .50$$

A. $\frac{1}{3} =$     $\frac{1}{8} =$     $\frac{2}{5} =$     $\frac{5}{6} =$

B. $\frac{7}{8} =$     $\frac{1}{6} =$     $\frac{2}{3} =$     $\frac{1}{5} =$

C. $\frac{3}{4} =$     $\frac{5}{8} =$     $\frac{7}{16} =$     $\frac{1}{16} =$

D. $\frac{3}{8} =$     $\frac{3}{5} =$     $\frac{4}{5} =$     $\frac{9}{16} =$

Sometimes, when you change fractions to decimals, the remainder may show a pattern.

$$\frac{4}{11} = 11\overline{)4.0000} \qquad \frac{4}{11} = .\overline{36}$$
$$\phantom{\frac{4}{11} = 11)}\underline{33}$$
$$\phantom{\frac{4}{11} = 11)}70$$
$$\phantom{\frac{4}{11} = 11)}\underline{66}$$
$$\phantom{\frac{4}{11} = 11)}40$$
$$\phantom{\frac{4}{11} = 11)}\underline{33}$$
$$\phantom{\frac{4}{11} = 11)}70$$
$$\phantom{\frac{4}{11} = 11)}\underline{66}$$
$$\phantom{\frac{4}{11} = 11)}4$$

$\frac{4}{11} = .\overline{36}$  Draw a line to indicate the repeating numbers.

Find the pattern. Draw a line over the first set of repeating numbers.

E. $\frac{7}{11} =$     $\frac{2}{9} =$

F. $\frac{5}{12} =$     $\frac{2}{11} =$

G. $\frac{2}{15} =$     $\frac{7}{22} =$

H. $\frac{4}{9} =$     $\frac{1}{3} =$

© Milliken Publishing Company    39    MP3378

Name_____   Multiplying Decimals

## COUNT FROM RIGHT — ADD TO LEFT

When multiplying decimals, remember to count the decimal places from the right. You may have to write in zeros on the left, and then place the decimal point.

(A) .04 × .4    (B) 200 × .07    (C) 6.9 × .08    (D) .345 × .08    (E) 12 × .006

(F) 227 × .04    (G) .024 × 62    (H) 31.9 × 2.3    (I) 600 × .002    (J) 2.06 × 4.3

(K) 320 × 4.21    (L) .028 × .13    (M) .042 × 7.5    (N) .213 × 9.2    (O) 3.95 × .032

(P) 409 × 3.26    (Q) .37 × .056    (R) .027 × .89    (S) 4000 × .035    (T) .0038 × 41.2

(U) .001 × .05    (V) .056 × .55    (W) 534 × 6.5    (X) .243 × .04    (Y) .008 × .09

What do you call a rabbit that's mad at the sun? Match your answers and write the letters below.

____  ____  ____   ____  ____  ____   ____   ____   ____  ____  ____  ____  ____
73.37 .1264 .15656  .552 .02403 .1264  140    140    14  .00005 1.9596 1.9596 .00072

Name_____

Multiplying and Dividing
by Powers of Ten

## WATCH THAT POINT!

Divide by 10? Move one place to the left.
Divide by 100? Move two places to the left.
Divide by 1000? _____
Divide by 10,000? _____

Got the point? Try these.

A. 47.5 ÷ 10 =         6.5 ÷ 10 =         23.91 ÷ 1000 =

B. 47.5 ÷ 100 =        .39 ÷ 10 =         7.44 ÷ 100 =

C. 47.5 ÷ 1000 =       8.53 ÷ 100 =       5.32 ÷ 1000 =

D. 47.5 ÷ 10,000 =     .97 ÷ 1000 =       .014 ÷ 100 =

To multiply, move right.
Multiply by 10? Move one place to the right.
Multiply by 100? Move two places to the right.
Multiply by 1000? _____
Multiply.

E. 47.5 × 10 =         2.65 × 100 =       4.56 × 1000 =

F. 47.5 × 100 =        54.93 × 1000 =     3.449 × 10 =

G. 47.5 × 1000 =       7.612 × 10 =       .9763 × 100 =

H. 47.5 × 10,000 =     35.7 × 100 =       .931 × 10,000 =

Watch the signs!

I. 5.8 × 100 =         71.0 ÷ 100 =       5.32 × 10 =

J. 22.4 ÷ 100 =        1.026 × 1000 =     .43 ÷ 100 =

K. 52.8 ÷ 1000 =       1.5 × 10 =         12.4 ÷ 1000 =

L. .314 × 10 =         .434 ÷ 1000 =      89.05 ÷ 100 =

Name_____    Dividing Decimals

## MOVE THAT POINT!

Find the quotients. Remember to move the decimal point in both the divisor and the dividend.

A.  .4)0.28    .6)0.54    .8).056    .9).07218    .6).246

B.  1.8)1.134    1.9)1.71    .62).1550    2.4)3.60    .37).3589

C.  7.1)3.55    .68).0408    .58).3886    9.7).07954

D.  .68).0272    9.6)8.160    38)2.736    90)6.120

E.  .60).18    6.7)11.524    .48).00576    2.4).0324

The coach at Pea U. spent $17.55 for soap. Each bar cost $.39. How many bars did he buy? _____

It cost $.14 to wash each towel. Coach spent $11.90 for towel laundry. How many towels were laundered? _____

Name_____

Quotients and Remainders as Decimals

## SHOPPING FOR ANSWERS

Divide and express the remainders as decimals to the nearest tenth or hundredth.

A. 9).081        5)239        8)32.40        34)9.01

B. 4)1.98       5)176        3)23.4         4)1.06

C. .06)3.795    .6)1.95      13)282.1       .5)3.8

Now, compare prices.
Add zeros to the dividend and divide.
Which of the two products costs less per item?

D. 4 cans of beans @ $1.30 or 6 cans @ $1.92         _____

E. 3 cans of juice @ $.98 or 4 cans @ $1.29          _____

F. 2 pounds of fudge @ $4.59 or 6 pounds @ $13.76    _____

G. 2 pounds of carrots @ $0.69 on 5 pounds @ $1.71   _____

H. 10 oranges @ $1.32 on 12 oranges @ $1.62          _____

I. 5 pencils for $1.25 or 3 for $.69                 _____

J. 3 notebooks for $2.67 or 2 for $1.79              _____

© Milliken Publishing Company
MP3378

Name_____

Rounding Decimals in Quotients

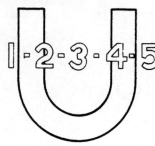 to read the instructions carefully!

To round a decimal quotient, you need:
 2 decimal places to round to the nearest tenth.
 3 decimal places to round to the nearest hundredth.
 4 decimal places to round to the nearest thousandth.

Round to the nearest tenth.

| First | | Then | |
|---|---|---|---|
| Divide to hundredths | $4\overline{)5.26}$ quotient $1.31$ (work shown) | Round the quotient to the tenths | 1.3 |

A.  $3\overline{)7.14}$ ____   $6\overline{)4.63}$ ____   $8\overline{)9.2}$ ____   $6\overline{).82}$ ____

B.  $46\overline{)29.7}$ ____   $18\overline{)12.78}$ ____   $21\overline{)9.48}$ ____   $57\overline{)6.84}$ ____

Round to the nearest hundredth.

C.  $5\overline{).17}$ ____   $4\overline{)2.14}$ ____   $7\overline{)1.36}$ ____   $9\overline{)6.2}$ ____

D.  $21\overline{)29.25}$ ____   $18\overline{)40.6}$ ____   $33\overline{)1.82}$ ____   $7\overline{).050}$ ____

Round to the nearest thousandth.

E.  $9\overline{)2.81}$ ____   $7\overline{).926}$ ____   $4\overline{).1451}$ ____   $6\overline{)42.83}$ ____

F.  $24\overline{)17.54}$ ____   $42\overline{)61.54}$ ____   $31\overline{)52.34}$ ____   $24\overline{)8.78}$ ____

**\*Counting on you…**

# Answer Key

## Page 1
6, 3, 8, 10, 12
1, 2, 4, 4, 3
3/8, 3/4, 5/6, 4/9
4/10, 1/3, 3/4
1/4, 1/2, 6/6 or 1

## Page 2
9 ÷ 3 = 3, 12 ÷ 4 = 3, 6 ÷ 3 = 2, 10 ÷ 2 = 5
1. 10 ÷ 2 = 5
2. 9 ÷ 3 = 3
3. 12 ÷ 6 = 2
4. 20 ÷ 5 = 4
5. 30 ÷ 6 = 5
6. 64 ÷ 8 = 8
7. 21 ÷ 7 = 3
8. 24 ÷ 12 = 12
9. 80 ÷ 2 = 40
10. 50 ÷ 10 = 5
11. 66 ÷ 3 = 22
12. 100 ÷ 10 = 10
13. 25 ÷ 5 = 5
14. 32 ÷ 8 = 4
15. 49 ÷ 7 = 7

Accept ⟌ format for division sentences.

## Page 3
1. 6/8—3/4
   5/10—1/2
   6/9—2/3
   4/12—1/3
   1/4—2/8

2. 4/10—2/5
   3/12—1/4
   8/16—1/2
   1/2—3/6
   2/14—1/7

3. 2/3—6/9
   5/15—1/3
   4/8—1/2
   1/5—3/15
   14/16—7/8

4. 2/5—4/10
   2/3—4/6
   18/36—1/2
   8/10—4/5
   3/4—12/16

## Page 4
Silo #1
1. >
2. <
3. >
4. <
5. <
6. =
7. >
8. =
9. <
10. <
11. =
12. =
13. <
14. <
15. >
16. =
17. >
18. >
19. <
20. <

## Page 5
1. 1 2/3
2. 1 2/8 or 1 1/4
3. 6/6 or 1 1/5
4. 4/3 or 1 1/3
5. 8/6 or 1 1/3
6. 6/5 or 1 1/5
7. 11/7 or 1 4/7
8. 3/3 or 1
9. 2 4/8 or 2 1/2
10. 6 3/10
11. 4 2/5
12. 1 7/8 or 2 5/8
13. 1 1/3 or 2 1/3
14. 2 5/10 or 2 3/5
15. 5 5/8 or 5 1/2
16. 7 19/10 or 8

Answers for 17 through 20 will vary.

## Page 6
.4, .2, .6, .8
3—.3, 7—.7, 9—.9, 5—.5
  A. .3, 2.1, 4.6, 6.4, .5
  B. 1.3, 7.7, 1.2, 4.1, .9
  C. 7/10, 1/5, 1 3/5, 9 1/5, 2/5
  D. 9/10, 5.1, 2.6, 1.8
  E. 1/2, 1.3, 3 1/10, .9

## Page 7
8 1/2, 1/2, 1 1/3, 1 2/3
5 1/4, 3 1/5, 1 2/5, 2 1/2
4 1/4, 1 1/10, 1/2, 5 5/8
1 1/2, 1, 2 4/5, 3 3/7
4 7/10, 1/4, 5 3/5, 1/2

## Page 8
1. 12—8/12, 3/12
2. 20—5/20, 4/20
3. 30—10/30, 18/30
4. 12—10/12, 3/12
5. 15—10/15, 9/15
6. 24—3/24, 16/24
7. 14—4/14, 7/14
8. 6—3/6, 5/6, 4/6
9. 10—2/10, 1/10, 2/10
10. 24—16/24, 2/24, 3/24
11. 6—2/6, 1/6, 3/6
12. 20—5/20, 10/20
13. 18—15/18, 3/18, 2/18
14. 12—4/12, 9/12, 2/12

## Page 9
1. 19/30
2. 7/8
3. 1/2
4. 1 1/10
5. 1 1/16
6. 8/9
7. 1 5/12
8. 9/10
9. 1
10. 1/2

1. 1/10
2. 1/4
3. 2/5
4. 1/6
5. 11/40
6. 1/6
7. 1/6
8. 3/10
9. 1 3/24
10. 1/3
11. 5/7
12. 2/9
13. 11/20
14. 1/5
15. 5/12

## Page 10
1. 1/3, 1/2, 3/4, 1/2
2. 2/3, 4/5, 4/5, 3/5, 1/2, 1/4
3. 1/4, 1, 1/2, 1/2
4. 4/5, 1, 31/50, 5/6
5. 1/3, 1/7, 1/5
6. 1/2, 1/5, 2/3

## Page 11
3/4
3/7
5/8          1/4
1/2          4/5
1/2          5/6
7/8          3/10
3/4          1/5
2/7          9/10
1/6          4/5
1/4          1/5

16/28—4/7
6/24—1/4
15/24—5/8
18/42—3/7
21/24—7/8
50/100—1/2

i

MP3378

**Page 12**

A. C, S, V
B. V , C, S
C. V, C, S, C, C, V
D. 2 1/8, 4 2/3, 1 6/9, 4 1/6, 1, 2, 3 1/7, 1 1/8
E. 9/2, 17/3, 33/5, 21/8, 43/7, 33/10, 40/6, 7/5

**Page 13**

1, 1 1/3, 1 1/4
1, 1 1/2, 1 1/2, 5/7
11/16, 3/5, 1 1/2, 4/5
1 1/3, 1 3/5, 3/4, 1, 1 1/7
9/10, 1 2/5, 5/8, 1
1 3/4, 3/4, 3/8, 1 1/3, 1 2/7

**Page 14**

| | | | |
|---|---|---|---|
| 1. 1/4 | 6. 1/9 | 11. 1/3 | 16. 1/3 |
| 2. 1/10 | 7. 1/4 | 12. 1/10 | 17. 2/5 |
| 3. 3/5 | 8. 0 | 13. 1/2 | 18. 1/2 |
| 4. 5/8 | 9. 2/9 | 14. 1/4 | 19. 1/3 |
| 5. 1/6 | 10. 3/10 | 15. 1/7 | 20. 2/5 |

**Page 15**

| | | |
|---|---|---|
| 1. 5/12 | 6. 11/15 | 11. 1 3/5 |
| 2. 1 1/6 | 7. 2/5 | 12. 1/3 |
| 3. 7/8 | 8. 3/4 | 13. 4/7 |
| 4. 1 1/10 | 9. 5/18 | 14. 1 |
| 5. 1 1/10 | 10. 3 | 15. 1 1/5 |

**Page 16**

A. 81 1/4, 7 5/6, 7 4/5, 6 3/7
B. 21 5/6, 7 3/7, 2 3/5, 4 1/2
C. 4 3/5, 2 5/8, 13/16, 1 13/14
D. 82 2/3, 9/10, 3 2/5, 1 3/4
E. 3 4/9, 17 3/4, 1 3/5, 2 2/7
F. 3 4/5, 9 11/12, 4 3/10, 9 3/4

**Page 17**

A. 4 2/7, 3 29/33, 8
B. 6 36/43, 5/12, 9/10
C. 7/50, 5/32, 5/8
D. 2 14/15, 9 1/2, 18
E. 2/25, 1/25, 18
F. 1 13/35, 5 5/16, 18/91
G. 7/26, 1 9/22, 2 4/15
H. 3 1/9, 2 6/13, 2 1/5
I. 4 1/7, 1 7/8, 7/18

**Page 18**

| | | | |
|---|---|---|---|
| 1. 2/7 | 6. 3/8 | 11. 5/12 | 16. 1/12 |
| 2. 3/20 | 7. 9/28 | 12. 1/4 | 17. 1/10 |
| 3. 4/15 | 8. 2/35 | 13. 1/4 | 18. 7/16 |
| 4. 1/10 | 9. 1/15 | 14. 1/6 | 19. 3/14 |
| 5. 26/45 | 10. 1/6 | 15. 1/5 | 20. 1/9 |

**Page 19**

| | | | |
|---|---|---|---|
| 1. 1 | 6. 8/81 | 11. 11 1/4 | 16. 64 |
| 2. 2/19 | 7. 25 | 12. 1/9 | 17. 4/17 |
| 3. 5 1/7 | 8. 2/7 | 13. 5/128 | 18. 2/7 |
| 4. 16 | 9. 1/10 | 14. 49 | 19. 1/16 |
| 5. 1 1/2 | 10. 1/6 | 15. 25 | 20. 4/9 |

**Page 20**

| | | | |
|---|---|---|---|
| 1. 4 1/2 | 8. 8 3/4 | 15. 69 3/10 | 22. 24 1/8 |
| 2. 4 3/8 | 9. 5 2/3 | 16. 84 3/8 | 23. 192 17/30 |
| 3. 13 | 10. 8 | 17. 32 | 24. 14 1/8 |
| 4. 8 4/5 | 11. 19 1/3 | 18. 25 9/10 | 25. 110 1/2 |
| 5. 3 3/7 | 12. 23 1/8 | 19. 27 5/9 | 26. 79 11/18 |
| 6. 10 2/7 | 13. 39 1/10 | 20. 5 3/10 | 27. 14 11/12 |
| 7. 10 1/2 | 14. 112 2/9 | 21. 14 3/4 | 28. 145 23/24 |

**Page 21**

A. 3 1/8 3 2/5 4 1/2 11 1/9
B. 2 1/4 5 13/16 39 1/2 3 1/2
C. 3 1/8 4 7/10 11 8/15 8 7/15
D. 14 1/36 9 1/3 78 9 3/8
E. 26 1/3 6 1/8 4 7/8 34 1/15
F. 14 7/30 33 5/14 65 11/45 48 19/30

**Page 22**

A. 4.3 2.8 24.07 .008
B. 16.32 .285 .0360 .5
C. 70/100 4/100 311/1000 14 5/1000
D. 8 2/10 24 8/10 4/10,000 3 15/100
E. .346 3.06
F. 40.48 .0606
G. 1.74 .0680
H. .2 .23 .0040
I. 4.12 30.2 .0024

**Page 23**

A. .36 1.36 1.205 .1210 6.44
B. 4.32 5.460 12.12 44.668 3.752
C. .006 .32 .096 .0030 .09
D. .18 .00040 .040 .00280 .10
E. .0032 .084 .024 .00004 .024

**Page 24**

A. 6.3 2 8
B. 2 11.5 3
C. 2 2 1.110891
D. 1 1/2 15/22 2
E. 56/115 1 7/28 1 1/99
F. 18/35 1/4 7/15
G. 1 5/22 3/10 5 15/16

**Page 25**

.1 .5 .2 .04
.10 .6 .20 .9
.3 .5

| | | |
|---|---|---|
| 1. .50 | 4. .3 | 7. .1 |
| 2. .3 | 5. .8 | 8. .60 |
| 3. .30 | 6. .5 | 9. .75 |

**Page 26**

1 2
1 2

| | |
|---|---|
| .8 | .17 |
| 1.8 | 37.4 |
| 2.4 | .91 |
| .6 | 1.1 |
| 8.29 | .7 |
| 7.8 | 2.8 |
| 2.04 | .5 |
| 5.38 | 1.92 |
| .9 | 9.80 |
| .85 | 1.23 |

seven and eight tenths
one and five hundredths
twenty-one and six tenths
fifteen
five and six tenths
one and sixty hundredths
seven hundredths
six and six tenths

**Page 26 (con't.)**
two tenths
one hundred forty-two and forty-two hundredths
one and thirty-eight hundredths
one and thirty-eight hundredths
three and nine tenths
eight and eight hundredths
sixty-eight hundredths
ten hundredths
seven and nine tenths
four tenths
fifty hundredths
five
five tenths

**Page 27**
A. <  <  <
B. <  <  <
C. <  <  >
D. <  >  >
LETTERS
E. 4.3, 5.7, 6.2
F. 5.1, 5.12, 51.2
G. 7.34, 7.36, 7.4
H. 4.2, 4.21, 42
I. (from top)
3.21, 3.31, 3.33, 3.41, 3.62

**Page 28**
A. 5.1      5.66      49.40
B. 39.01    9.81      40.19     12.6
C. 26.17    25.45     129.98    177.32    41.84
D. 65.12    25.15     186.97    30.16     153.23
E. 3.73 miles

**Page 29**
A. 2.2      5.52      6.24
B. .15      22.24     22.86     3.38
C. 1.0      4.69      57.97     62.78     23.63
D. 37.9     4.68      6.5       3.36      .66
E. 3.82     .37       .98       4.79      4.2
F. 2.4      1.12      19.5      3.3
G. 1.2      1.7       21.19     1.98

**Page 30**
A. $3.25    $5.83     $3.66     $4.86     $7.38
B. $2.42    $2.34     $1.91     $3.20     $4.44
C. $13.21
D. $29.20
E. $75.90
F. $2.86

**Page 31**
A. .42–2.042
B. .3–.614
C. .32–6.041
D. .561–40.002
E. .423–7.069
F. .042–4.476
G. .002–87.087
H. seven tenths–five thousandths
I. zero thousandths–zero tenths
J. one thousandth–two thousandths
K. seven hundredths–six hundredths
DECIMALS

**Page 32**
A. .8 / .04
B. .08 / .4
C. .80 / .40
D. $60/100 – 3/10$
E. $3/100 – 30/100$
F. $5/10 – 3/100$
G. .45 / .11 / 7.30 / 12.10
H. .60 / .15 / 5.60 / 28.40
I. .05 / .13 / 1.40 / 3.04
J. .29 / .20 / 8.04 / 9.03
K. .13 / .01 / 6.05 / 9.30
L. $1/5, 3/4, 9/10, 1/20$
M. $1/4, 3/10, 3/25, 7/20$
N. $6¼, 2½, 6⅕, 3¼$
O. $37/100, 31/50, 7 7/10, 4 2/25$
P. $7/25, 5¾, 9/20, 3⅗$

**Page 33**
A. 45.461    30.380    31.847
B. .1552     9.403     56.196    1.3711    7.596
C. 2.6257    .922      4.505     .8509     24.181
D. 14.217    7.0917    2.9891    1058.733
E. .3189     1.713     .0575     .191

**Page 34**
A. $46.90–47, $41.90–42, $17.99–18, $33.00–33
B. $37.16–37, $59.21–59, $1195.21–1195, $42.19–43
C. $63.10–63, $52.27–52, $17.39–18, $29.08–30
D. No–$6.00
E. Yes–$10.00

**Page 35**
A. 3.0       56.5      23.4      24.8      379.4
B. 43.8      114.8     67.2      24.06     22.029
C. 178.08    125.02    8.874     144.48    64.600
D. 17.440    134.4     26.250    0.630     4.32
E. 261.38    8.51      577.20    1496.5    315.27

5338–BEES

**Page 36**
A. .35526–.36, 14.72–15
B. 3.690–4.0, 10.92–12, .3456–.35, .918–1.0
C. .06853–.072, .4824–.48, 22.0524–22, .18492–21
D. 1.6490–1.5, .50484–.48, .01813–.020, 91.2374–88

**Page 37**
A. .29       9.1       .4        .4        9.7
B. 1.4       .53       10.4      .71       3.13
C. .26       .6        .368      .2        .024
D. .302      .36       .8        .24       4.1
E. .406      .327      .802      .250      2.34

6.52 inches per month
3.99 hours

**Page 38**
A. <, >
B. <, <
C. >, =, <, =
D. <, >, <, <
E. >, >, =, <

(F) Worm(G)s are (H)in ap(I)ples!

**Page 39**
| | | | | | | |
|---|---|---|---|---|---|---|
| A. .33 | .13 | .40 | .83 | E. .6363 | .2222 | |
| B. .88 | .17 | .67 | .20 | F. .4166 | .1818 | |
| C. .75 | .63 | .44 | .06 | G. .1333 | .31818 | |
| D. .38 | .60 | .80 | .56 | H. .4444 | .3333 | |

**Page 40**
A. .016   B. 14.00   C. .552   D. .02760   E. .072
F. 9.08   G. 1.488   H. 73.37   I. 1.200   J. 8.858
K. 1347.20   L. .00364   M. .3150   N. 1.9596   O. .12640
P. 1333.34   Q. .02072   R. .02403   S. 140.000   T. .15656
U. .00005   V. .03080   W. 3471.0   X. .00972   Y. .00072
HOT CROSS BUNNY

**Page 41**
3 places to the left
4 places to the left

A. 4.75 / .65 / .02391
B. .475 / .039 / .0744
C. .0475 / .0853 / .00532
D. .000475 / .00097 / 00014

3 places to the right

E. 475.0 / 265.00 / 4560.00
F. 4750.0 / 54,930.00 / 34.490
G. 47,500.0 / 76.120 / 97.6300

H. 475,000.0 / 3570.0 / 9310.000
I. 580.0 / .71 / 53.20
J. .224 / 1026 / .0043
K. .0528 / 15.0 / .0124
L. 3.140 / .000434 / .8905

**Page 42**
A. .7   .9   .07   .0802   .41   D. .04   .85   .072   .068
B. .63   .9   .25   1.5   .97   E. .3   1.72   .012   .0135
C. .5   .06   .67   .0082

45 bars
85 towels

**Page 43**
A. .009 / 47.8 / 4.05 / 0.27    F. 6 @ $13.76
B. 0.50 / 35.2 / 7.8 / 0.27     G. 5 @ $1.71
C. 63.25 / 3.25 / 21.7 / 7.6    H. 10 @ $1.32
D. 6 @ $1.92                    I. 3 for 69¢
E. 4 @ $1.29                    J. 3 for $2.67

**Page 44**
A. 2.38–2.4 / .77–.8 / 1.15–1.2 / .13–.1
B. .64–.6 / .71–.7 / .45–.5 / .12–.1
C. .034–.03 / .535–.54 / .194–.19 / .688–.69
D. 1.392–1.39 / 2.255–2.26 / .055–.06 / .007–.01
E. .3122–.312 / .1322–.132 / .0362–.036 / 7.1383–7.138